United States
Environmental Protection
Agency

Office of Water

Climate Change Adaptation Implementation Plan

Report Number: EPA-100-K-14-001A

May 2014

Prepared by:
Office of Water
U.S. Environmental Protection Agency
May 2014

Disclaimer

To the extent this document mentions or discusses statutory or regulatory authority, it does so for informational purposes only. This document does not substitute for those statutes or regulations, and readers should consult the statutes or regulations to learn what they require. Neither this document, nor any part of it, is itself a rule or a regulation. Thus, it cannot change or impose legally binding requirements on EPA, States, the public, or the regulated community. Further, any expressed intention, suggestion or recommendation does not impose any legally binding requirements on EPA, States, tribes, the public, or the regulated community. Agency decision makers remain free to exercise their discretion in choosing to implement the actions described in this Plan. Such implementation is contingent upon availability of resources and is subject to change.

Preface

The U.S. Environmental Protection Agency (EPA) is committed to identifying and responding to the challenges that a changing climate poses to human health and the environment.

Scientific evidence demonstrates that the climate is changing at an increasingly rapid rate, outside the range to which society has adapted in the past. These changes can pose significant challenges to the EPA's ability to fulfill its mission. The EPA must adapt to climate change if it is to continue fulfilling its statutory, regulatory and programmatic requirements. The Agency is therefore anticipating and planning for future changes in climate to ensure it continues to fulfill its mission of protecting human health and the environment even as the climate changes.

In February 2013, the EPA released its draft *Climate Change Adaptation Plan* to the public for review and comment. The plan relies on peer-reviewed scientific information and expert judgment to identify vulnerabilities to EPA's mission and goals from climate change. The plan also presents 10 priority actions that EPA will take to ensure that its programs, policies, rules, and operations will remain effective under future climatic conditions. The priority placed on mainstreaming climate adaptation within EPA complements efforts to encourage and mainstream adaptation planning across the entire federal government.

Following completion of the draft *Climate Change Adaptation Plan*, each EPA National Environmental Program Office, all 10 Regional Offices, and several National Support Offices developed a *Climate Adaptation Implementation Plan* to provide more detail on how it will carry out the work called for in the agency-wide plan. Each *Implementation Plan* articulates how the office will integrate climate adaptation into its planning and work in a manner consistent and compatible with its goals and objectives.

Taken together, the *Implementation Plans* demonstrate how the EPA will attain the 10 agency-wide priorities presented in the *Climate Change Adaptation Plan*. A central element of all of EPA's plans is to build and strengthen its adaptive capacity and work with its partners to build capacity in states, tribes, and local communities. EPA will empower its staff and partners by increasing their awareness of ways that climate change may affect their ability to implement effective programs, and by providing them with the necessary data, information, and tools to integrate climate adaptation into their work.

Each Program and Regional Office's *Implementation Plan* contains an initial assessment of the implications of climate change for the organization's goals and objectives. These "program vulnerability assessments" are living documents that will be updated as needed to account for new knowledge, data, and scientific evidence about the impacts of climate change on EPA's mission. The plan then identifies specific priority actions that the office will take to begin addressing its vulnerabilities and mainstreaming climate change adaptation into its activities. Criteria for the selection of priorities are discussed. An emphasis is placed on protecting the most vulnerable people and places, on supporting the development of adaptive capacity in the tribes, and on identifying clear steps for ongoing collaboration with tribal governments.

Because EPA's Programs and Regions and partners will be learning by experience as they mainstream climate adaptation planning into their activities, it will be essential to evaluate their efforts in order to understand how well different approaches work and how they can be improved. Each *Implementation Plan* therefore includes a discussion of how the organization will regularly evaluate the effectiveness of its adaptation efforts and make adjustments where necessary.

The set of *Implementation Plans* are a sign of EPA's leadership and commitment to help build the nation's adaptive capacity that is so vital to the goal of protecting human health and the environment. Working with its partners, the Agency will help promote a healthy and prosperous nation that is resilient to a changing climate.

Bob Perciasepe

Deputy Administrator

September 2013

Table of Contents

Appendices

Page intentionally left blank

INTRODUCTION

This *Implementation Plan* provides an overview of the opportunities available to the Office of Water within the U.S. Environmental Protection Agency (EPA) to respond to the challenges that a changing climate poses for the successful operation of national programs to protect the quality of the national water resources and drinking water.

This *Plan* is organized using the framework adopted by the EPA for each of its major national offices and regional offices. Key elements of the *Plan* address:

> ➢ Vulnerability of water resources, including clean water and drinking water programs, to climate change;
> ➢ Priority actions for water program response to climate change;
> ➢ Office of Water contribution to meeting EPA strategic measures on climate change;
> ➢ Legal and enforcement issues;
> ➢ Training and outreach for climate change adaptation;
> ➢ Partnerships with Tribes;
> ➢ Populations and places vulnerable to a changing climate; and
> ➢ Program evaluation and cross-Agency pilot projects.

This *Plan* draws on, and is intended to help implement, the EPA *National Water Program 2012 Strategy: Response to Climate Change* published in December 2012. *The 2012 Strategy* describes long-term goals for the management of sustainable water resources in light of climate change and is intended to be a roadmap to guide future programmatic planning and inform decision makers during the Agency's annual planning process. The *2012 Strategy* is available at http://www.epa.gov/water/climatechange. Some initial steps by EPA national water programs and regional offices are described in *the 2012 Highlights of Progress* Report (available at the website above).

> "Many of the programs and activities already underway throughout the National Water Program...are even more important to do in light of climate change. However, climate change poses such significant challenges to the nation's water resources, that more **transformative approaches will be necessary**."
>
> *National Water Program 2012 Strategy: Response to Climate Change*; EPA; 2012; p. 1

The *2012 Strategy* identifies five long-term programmatic vision areas:

1. water infrastructure;
2. watersheds and wetlands;
3. coastal and ocean waters;
4. water quality; and
5. working with Tribes.

Each of these programmatic vision areas is supported by more specific goals and strategic actions. Additional goals and actions address "cross-cutting program support" topics. A table providing a brief summary of the 5 vision areas, 19 goals, and 53 strategic actions described in the *2012 Strategy* is provided in Appendix 1. Climate change issues and actions in climate regions across the country are also described in the *2012 Strategy*.

> **Climate Change Vision:**
>
> The National Water Program's overarching vision for responding to climate change is:
>
> *Despite the ongoing effects of climate change, the National Water Program intends to continue to achieve its mission to protect and restore our waters so that drinking water is safe; and aquatic ecosystems sustain fish, plants, and wildlife, as well as economic, recreational, and subsistence activities.*

The National Water Program, including both EPA Headquarters offices and EPA Regional offices, has taken several steps to implement the new *2012 Strategy* including:

➢ Developing this *Implementation Plan* as part of the Agency-wide work to prepare EPA climate change adaptation implementation plans;

➢ Preparing internal workplans for 2013 and 2014 describing specific implementation actions that EPA Headquarters and Regional offices plan to implement;

➢ Committing to the continued operation of the EPA National Water Program State and Tribal Climate Change Council made up of representatives of States and Tribes to provide advice and guidance to the National Water Program in addressing issues related to climate change and water;

➢ Identifying future directions for the work on climate change issues in the EPA *FY 2015 Addendum* to the EPA *FY 2014 National Water Program Guidance for FY 2014;* and

➢ Committing to leadership of cooperative efforts with other organizations to address climate change and water issues including serving as co-chair of the Interagency Climate Change and Water Workgroup (made up of Federal agencies) and the Climate Change Workgroup of the Advisory Committee on Water Information (ACWI) made up of 40 stakeholder organizations and Federal agencies.

A graphic illustrating the organizational framework for the climate change work by the National Water Program is provided in Appendix 2. Appendix 3 provides a list of the principal members of the National Water Program Climate Change Workgroup that developed this *Plan.*

II. Vulnerability of Water Resources to Climate Change

The many impacts that a changing climate is likely to have on water resources, both freshwater and coastal resources, include:

1. Increases in water pollution problems as air and water warm;
2. More extreme weather events;
3. Changes to water availability (rain and snow level and distribution);
4. Sea level rise/storm surge and waterbody boundary movement and displacement;
5. Collective impacts on coastal areas; and
6. Indirect impacts resulting from changes in energy and fuel production.

The nature and extent of these impacts is described in greater detail below, together with the relationship of these impacts to clean water and drinking water program components. The association of climate change impacts on water with clean water and drinking water programs is illustrated using a chart of the full range of programs (see sample chart below). For each of the six climate change impacts identified above, the specific programs identified in the chart affected by the impacts are highlighted.   This chart format illustrates both the programs expected to be affected by the specific type of climate change impacts and the programs expected to be less affected.

Clean Water and Drinking Water Program Template

Drinking Water Standards	Surface Water Standards	Technology Based Standards	Emergency Planning
Drinking Water Planning	Clean Water Planning	Water Monitoring	Water Restoration/ Total Maximum Daily Loads (TMDLs)
Underground Injection Control Permits	Discharge Permits	Storm Water Permits	Wetlands Permits
Source Water Protection	Nonpoint Pollution Control	Coastal Zone	National Estuaries Program
Drinking Water SRF (State Revolving Funds)	Clean Water SRF (State Revolving Funds)	Ocean Protection	Combined Sewer Overflow Plans

A table illustrating the collective climate change impacts on clean water and drinking water program components is provided in Appendix 4.

1) Increases in Water Pollution Problems as Air and Water Warm:

Warmer air temperatures will result in warmer water. Warmer water holds less dissolved oxygen making instances of low oxygen levels or "hypoxia" more likely (which is detrimental to the aquatic ecosystem); foster harmful algal blooms; and alter the toxicity of some pollutants. For further details on these impacts see pages 41-51, 56, 81, 87, 96, 122, 150, 151 in Karl et al., 2009.

As air and water warm, water resource managers will likely face significant challenges:

Vulnerability Citations:
Descriptions of vulnerabilities provided in this section are drawn from: *Global Climate Change Impacts in the United States,* Thomas R. Karl, Jerry M. Melillo, and Thomas C. Peterson, (eds.) Cambridge University Press, 2009 (Karl et al 2009). Connections of climate impacts to water programs are the assessment of water program managers.

- increased pollutant concentrations and lower dissolved oxygen levels will result in higher incidences of impaired water quality; (Karl et al.; p. 46)
- increased growth of algae and microbes will affect drinking water quality; (Ibid; p. 46, 96)
- some aquatic taxa will be replaced by other taxa better adapted to warmer water (i.e., cold water fish will be replaced by cool water fish), and this process will likely occur at an uneven pace disrupting aquatic system health and allowing non-native and/or invasive species to become established; (Ibid; p. 81, 87, 122)
- warmer air temperatures will increase demand for cooling and for power production, resulting in increased discharges of warm water from power plants; (Ibid; p. 49, 56)
- increased water use will put stress on water infrastructure and demands on the clean water and drinking water State Revolving Funds; (Ibid; p. 48, 49) and
- increased evapotranspiration rates resulting from temperature increases may result in water losses for which drinking water and wetlands managers will need to account (Ibid; p. 49).

Pollution Problems Related to Warmer Air and Water: Effects on Water Programs
(Shaded areas reflect programs most affected by air and water temperature increases)

Drinking Water Standards	Surface Water Standards	Technology Based Standards	Emergency Planning
Drinking Water Planning	Clean Water Planning	Water Monitoring	Water Restoration/ TMDLs
Underground Injection Control Permits	Discharge Permits	Storm Water Permits	Wetlands Permits
Source Water Protection	Nonpoint Pollution Control	Coastal Zone	National Estuaries Program
Drinking Water SRF	Clean Water SRF	Ocean Protection	Combined Sewer Overflow Plans

2) More Extreme Weather Events:

Heavier precipitation from tropical and inland storms will increase flood risk, expand flood hazard areas, increase the variability of stream flows (i.e., higher high flows and lower low flows) and increase the velocity of water during high flow periods, thereby increasing erosion. These changes will have adverse effects on water quality, drinking water sources, and aquatic ecosystem health. For example, increases in intense rainfall may result in more nutrients, sediments, pathogens, and toxins being washed into waterbodies. For further details on these impacts, see pages 18, 25, 34-36, 41, 44, 88, 94, 95, 102 in Karl et al., 2009.

Water resource managers will face significant challenges as storm intensity increases:
- although there is uncertainty with respect to climate models addressing storm intensity and frequency, emergency plans for drinking water and wastewater infrastructure need to recognize the possibility of increased risk of high flow and high velocity events due to intense storms and potential low flow periods; (Ibid; p. 25, 34-36, 44)
- damage from intense storms may increase the demand for public infrastructure funding and may require re-prioritizing of infrastructure projects; (Ibid; p. 47)
- floodplains may expand along major rivers requiring protection or relocation of drinking water and wastewater infrastructure facilities and coordination with local planning efforts; (Ibid; p. 44)
- in urban areas, stormwater collection and management systems may need to be redesigned to increase capacity; (Ibid; p. 47, 48)
- combined storm and sanitary sewer systems may need to be redesigned because an increase in storm event frequency and intensity can result in more combined sewer overflows causing increased pollutant and pathogen loading to receiving waterbodies; (Ibid; p. 48, 94, 95)
- changes in precipitation patterns and intensity may increase the demand for watershed management that mitigates the impacts of intense storms and build resilience into water management through increased water retention (e.g., green infrastructure, smart growth, and source water protection practices); (Ibid; p. 41, 44, 45) and
- the management of wetlands for stormwater control purposes and to buffer the impacts of intense storms will be increasingly important. (Ibid; p. 88, 102)

Extreme Weather: Effects on Water Programs
(Shaded areas reflect programs most affected by storm intensity)

Drinking Water Standards	Surface Water Standards	Technology Based Standards	Emergency Planning
Drinking Water Planning	Clean Water Planning	Water Monitoring	Water Restoration/ TMDLs
Underground Injection Control Permits	Discharge Permits	Storm Water Permits	Wetlands Permits
Source Water Protection	Nonpoint Pollution Control	Coastal Zone	National Estuaries Program
Drinking Water SRF	Clean Water SRF	Ocean Protection	Combined Sewer Overflow Plans

3) Changes to Water Availability (Rain and Snow Level and Distribution):

In some parts of the country, droughts, changing patterns of precipitation and snowmelt, and increased water loss due to evaporation as a result of warmer temperatures, will result in changes to the availability of water for drinking and for use for agriculture, industry, and energy production, as well as fire protection. In other areas, sea level rise and salt water intrusion will have the same effect. Warmer air temperatures may also result in increased demands on water supplies, and the water needs for agriculture, industry, and energy production are likely to increase. For further details on these impacts see pages 44-46, 48, 82, 94, 95, 113, 124, 129 in Karl et al., 2009.

Changing precipitation patterns pose several challenges for water program managers:
- increased rainfall, especially more intense rainfall, will result in increased stormwater runoff and may make overflows of sanitary sewers and combined sewers more frequent, putting new demands on discharge permit and nonpoint pollution programs; (Ibid; p. 44, 45, 48, 94, 95)
- increased storm water runoff will wash sediment and other contaminants into drinking water sources, requiring additional treatment; (Ibid; p. 44-46)
- additional investments in water infrastructure may be needed to manage both decreases and increases in rainfall and these demands could increase demand for water financing generally, including from the State Revolving Funds; (Ibid; p. 44-46)
- limited water availability and drought in some regions will require drinking water providers to reassess supply facility plans and consider alternative pricing, allocation, water conservation, and water reuse options; (Ibid; p. 44, 48, 113, 129)
- in areas with less precipitation or reduced snowpack, demand for water may shift to underground aquifers and prompt water recycling and reuse, development of new reservoirs, or underground injection of treated water for storage; (Ibid; p. 44, 45, 124)
- in areas with less precipitation, reduced stream flow may make meeting water quality goals more challenging; (Ibid; p. 44, 45, 46) and
- increased incidence of wildfire as a result of higher temperatures and drought may increase soil erosion and sedimentation, increase water pollution, increase risk of flooding, and pose a threat to aquatic habitats and water infrastructure. (Ibid; p. 82)

Changes in Rainfall and Snowfall Levels/Distribution: Effects on Water Programs
(Shaded areas reflect programs most affected by rainfall and snowfall levels)

Drinking Water Standards	Surface Water Standards	Technology Based Standards	Emergency Planning
Drinking Water Planning	Clean Water Planning	Water Monitoring	Water Restoration/ TMDLs
Underground Injection Control Permits	Discharge Permits	Storm Water Permits	Wetlands Permits
Source Water Protection	Nonpoint Pollution Control	Coastal Zone	National Estuaries Program
Drinking Water SRF	Clean Water SRF	Ocean Protection	Combined Sewer Overflow Plans

4) Sea Level Rise/Storm Surge and Waterbody Boundary Movement and Displacement:

Rising sea levels will move ocean and estuarine shorelines by inundating lowlands, displacing wetlands, and altering the tidal range in rivers and bays. Storm surges resulting from more extreme weather events will increase the areas subject to periodic inundation. Changing water flow to lakes and streams, increased evaporation, and changed precipitation in some areas will affect the extent of wetlands and lakes. Water levels in the Great Lakes are expected to fall. For further details on these impacts, see pages 12, 25, 34, 47, 48, 84, 109, 114, 138, 152 in Karl et al., 2009.

Sea level rise, storm surges, and waterbody movement will affect a range of water programs and pose significant challenges for water program managers:

- emergency plans for drinking water and wastewater infrastructure need to account for long-term projections for rising sea levels; (Ibid; p. 12, 25, 34, 47, 114)
- drinking water systems will need to consider relocating intakes as sea levels rise and salt water intrudes into freshwater aquifers used for drinking water supply and protecting or relocating some treatment facilities; (Ibid; p. 47, 138)
- wastewater utilities will need to consider hardening facilities against storm surge, protecting facilities with natural or man-made barriers, and relocation of some treatment facilities and discharge outfalls as a result of sea level rise; (Ibid; p. 12, 109, 152) and
- watershed-level planning will need to incorporate an integrated approach to coastal management in light of sea level rise including land use planning, building codes, land acquisition and easements, shoreline protection structures (e.g., seawalls and channels), beach nourishment, wetlands management, and underground injection to control salt water intrusion to fresh water supplies. (Ibid; p. 12, 49, 84, 114)

Sea Level Rise/Storm Surge: Effects on Water Programs
(Shaded areas reflect programs most affected by sea level rise)

Drinking Water Standards	Surface Water Standards	Technology Based Standards	Emergency Planning
Drinking Water Planning	Clean Water Planning	Water Monitoring	Water Restoration/ TMDLs
Underground Injection Control Permits	Discharge Permits	Storm Water Permits	Wetlands Permits
Source Water Protection	Nonpoint Pollution Control	Coastal Zone	National Estuaries Program
Drinking Water SRF	Clean Water SRF	Ocean Protection	Combined Sewer Overflow Plans

5) Collective Impacts on Coastal Areas:

Coastal areas are likely to see multiple impacts associated with climate change including: sea level rise, increased damage from floods and storm surges, coastal erosion, changes in drinking water supplies, and increasing temperature and acidification of the oceans (e.g., decreases in pH, decreases in carbonate ion availability for calcifying organisms). These overlapping impacts make protecting water resources in coastal areas especially challenging. For further details on these impacts, see pages 17, 43, 47, 84, 85, 148, 151 in Karl et al., 2009.

Changes in ocean characteristics pose several challenges for water program managers including:

- watershed-level protection programs may need to be revised to account for changes in natural systems as salinity and pH levels change; (Ibid; p. 17, 43, 47, 151)
- climate change and ocean acidification may exacerbate existing stresses on coral reefs such that programs to protect coral reefs, including temperate and cold water corals, from land-based pollution and impacts may need to be reassessed to provide enhanced protection; (Ibid; p. 84, 85, 148) and
- wetlands programs may need to be adjusted to account for changing salinity levels and impacts on wetlands health. (Ibid; p. 47, 84)

Changing Ocean Characteristics: Effects on Water Programs
(Shaded areas reflect programs most affected by changing ocean characteristics)

Drinking Water Standards	Surface Water Standards	Technology Based Standards	Emergency Planning
Drinking Water Planning	Clean Water Planning	Water Monitoring	Water Restoration/ TMDLs
Underground Injection Control Permits	Discharge Permits	Storm Water Permits	Wetlands Permits
Source Water Protection	Nonpoint Pollution Control	Coastal Zone	National Estuaries Program
Drinking Water SRF	Clean Water SRF	Ocean Protection	Combined Sewer Overflow Plans

6) Indirect Impacts Resulting from Changes in Energy and Fuel Production:

Possible responses to climate change include development of alternative methods of energy and fuel production that reduce emissions of greenhouse gases, as well as development of carbon sequestration technology. Alternative methods of both energy production and sequestration can have impacts on water resources including: increased water use and withdrawals from surface waters and groundwater, potential nonpoint pollution impacts of expanded agricultural production, increased water temperatures due to discharge of process cooling waters and reduced assimilative capacity of warmer waters, increased pollution concentration due to low flows, and effects of carbon sequestration on groundwater or ocean environments. For further details on these impacts, see pages 48, 49, 56, 59, 60, 74 in Karl et al., 2009.

Changing energy generation methods poses several challenges for water program managers including:

- increased water use and withdrawals will require expanded efforts to ensure water supply availability; (Ibid; p. 48, 49, 56)
- increased water supply demands and more variable water supplies will have effects on water resource management and reservoir operation; (Ibid; p. 48, 49, 59, 60) and
- need for increased attention to discharge permit conditions to address increased temperature and concentration of pollutants due to low flows. (Ibid; p. 49, 56)

Energy Generation Shifts: Effects on Water Programs
(Shaded areas reflect programs most affected by energy generation shifts)

Drinking Water Standards	Surface Water Standards	Technology Based Standards	Emergency Planning
Drinking Water Planning	Clean Water Planning	Water Monitoring	Water Restoration/ TMDLs
Underground Injection Control Permits	Discharge Permits	Storm Water Permits	Wetlands Permits
Source Water Protection	Nonpoint Pollution Control	Coastal Zone	National Estuaries Program
Drinking Water SRF	Clean Water SRF	Ocean Protection	Combined Sewer Overflow Plans

III.   Priority Actions

Recognizing the impacts that a changing climate is likely to have on water resources generally, and clean water and drinking water programs more specifically, the Office of Water identified criteria to consider in defining "priority actions" to respond to these impacts and make the water programs more "climate ready." These criteria, and the "priority actions" defined based on the criteria, are described below.   The Office of Water commitment to "mainstreaming" climate considerations into the full range of core clean water and drinking water programs is also described. Some specific core programs that, although designed for other purposes, have clear benefits in responding to a changing climate, are identified.

In addition, EPA Regional offices play a central role in implementing clean water and drinking water programs and are especially important because they are in a position to tailor implementation actions to the varied climate change and water adaptation challenges that exist across the country. Additionally, EPA Regional offices are able to work closely with States, Tribes, and other stakeholders to advance these activities. The Office of Water and Regional office water programs have identified a set of nine "common" activities that water programs in each Regional office will attempt to implement to build their capability to respond to climate change challenges related to water resources. These common climate change and water actions are described at the end of this section.

Criteria to Identify Priority Actions

Since resources to implement climate change response actions are limited, it is important to consider the significance of the impacts and to allocate scarce resources to response actions that address the most pressing and critical threats.

Some key criteria to consider when linking climate change impacts to potential response actions include:

- ➢ **Urgency:** What is the timing of the impact? How urgent is it that it be addressed?
- ➢ **Risk:** How significant is the risk to public health, infrastructure, or aquatic ecosystems?
- ➢ **Geographic Scale:** What is the geographic/demographic scale of the impact?
- ➢ **Programmatic Scale:** What is the scale of the programmatic impact?
- ➢ **Probability of Occurrence:** What is the likelihood the impact will actually occur?

Office of Water Priority Actions

Based on consideration of the criteria identified above, the Office of Water has identified ten "priority actions". These actions were identified from a larger group of national program office and regional water program actions developed as part internal work planning processes.

1. Encourage water utilities to use the **Climate Resiliency Evaluation and Awareness Tool** (CREAT) and develop Version 3.0 of this tool.

2. Promote use of an **Extreme Events Workshop Planner** designed to provide everything a water sector utility needs to plan, customize, and conduct a workshop focused on planning for extreme events including flooding, drought, sea level rise and storm surges, wildfire, and reduced snowpack.

3. In cooperation with EPA Regional offices, **expand the number of WaterSense partners** nationally and in each Region, with a goal of a 150 additional partners annually.

4. Encourage the **Clean Water and Drinking Water State Revolving Funds** to incorporate climate change considerations into their processes.

5. Encourage the National Estuary Programs and other watershed managers at the state and local level to use the **Watershed Climate Change Adaptation Planning Workbook** developed by the EPA National Estuary Program.

6. Complete development of initial screening criteria to identify **water and wastewater facilities on the Atlantic and Gulf Coasts that may be at risk of inundation** in the event of a storm surge comparable to Hurricane Sandy and work with utilities to address risks.

7. Complete **technical development of the Climate Change Extension within the Stormwater Calculator** and identify and implement enhancements to improve its utility.

8. Work with EPA Regional office counterparts to identify ways to **better integrate climate change considerations into water quality management planning projects and processes** and develop an initial report.

9. Draft a white paper providing information States and Tribes can use to **protect aquatic life from negative effects associated with alteration of hydrologic conditions**, including potential effects from climate change.

10. Engage key stakeholders in climate change adaptation work by continuing to support the EPA **State and Tribal Climate Change Council** that advises the National Water Program and engaging additional stakeholders using a range of forums and mechanisms.

It is important to note that a key objective of the *2012 Strategy* is to integrate climate change considerations and awareness into day-to-day management decisions for clean water and drinking water programs at national, regional, state, tribal, and local levels. The National Water program is facilitating this "mainstreaming" of climate change into core water programs by providing information and training to water program managers on climate change issues and prompting discussions of opportunities to recognize climate change in program management wherever possible. The Office of Water is also working with water programs in the ten EPA Regions to address climate change and water issues generally, as well as specific challenges that occur in each Region (see common EPA regional climate change actions below).

In addition to the specific "priority actions" identified above, the National Water Program conducts a range of programs that, although not designed to directly or uniquely address the impacts of a changing climate, make important contributions to making water resources more resilient to the impacts of a changing climate. Some examples of climate change-supporting programs include:

- Wetlands programs that help protect and restore wetlands that serve as sponges to retain water from more intense storm events, increased precipitation, and more rapid snowmelt;

- Stormwater permit programs that reduce pollution levels and the rate of runoff of rainfall in developed areas with large percentages of impervious surfaces and programs that promote improved stormwater management through implementation of "green infrastructure" practices;

- Healthy watershed programs that help maintain the quality of healthy watersheds and supportive habitat corridor networks across the country that provide resilience to climate change impacts;

- The National Estuary Program supports development and implementation of Comprehensive Conservation and Management Plans (CCMPs) in 28 estuaries around the country and had supported the development of a range of projects to address climate change challenges faced in these estuaries;

- Infrastructure management programs to expand the use of management practices that make water and wastewater treatment facilities more sustainable, including practices that improve resilience to climate change; and

- Monitoring programs, such as the *National Coastal Condition Report*, that can provide benchmarks of progress in addressing key climate change impacts.

Common Climate Change Actions for Regional Water Programs

Each regional water program will attempt to carry out the following common climate change related activities in 2013:

1. **Participate in the National Water Program Climate Change Workgroup:** Maintain current participation in the National Water Program Climate Change Workgroup, including identifying a single point of contact for the regional water program.

2. **Support the EPA Office of Water Climate Change Adaptation Implementation Plan:** Help to develop and implement the EPA-wide Climate Change Adaptation Implementation Plan (and coordinate between the National Water Program *2012 Strategy* and the EPA regional climate change adaptation implementation plans.

3. **Build Internal Climate Change Communications:** After the completion of the EPA Climate Change Adaptation Implementation Plan, draw on materials developed by the Office of Water to provide training for regional water program staff on the challenges that climate change poses for water programs and familiarize them with the National Water Program *2012 Strategy* and regional climate adaptation plans through a variety of means such as "all hands" meetings, webinars, seminars, and dissemination of the plans.

4. **Build External Climate Change Communications:** Support national program efforts to inform and educate water program managers in the public and private sectors on climate change and water issues through a variety of means such as identifying key stakeholders and expanding professional networks, improving educational outreach efforts on national and regional EPA climate change websites and in other media, and disseminating clear and credible messaging on climate change science and impacts.

5. **Address Climate Change in Meetings with States and Tribes:** In program meetings with States and Tribes in 2014, include discussion of ongoing Agency and regional climate change adaptation planning, the National Water Program *2012 Strategy*, and climate change activities related to State water programs as appropriate.

6. **Support Coordination among Federal Agency Regional Offices:** Coordinate with the Regional offices of other Federal agencies on climate change adaptation matters and participate, where appropriate, with related interagency cooperative and collaborative efforts to address climate change challenges on a regional scale.

7. **Promote Use of Tools from the Climate Ready Water Utilities (CRWU) and Climate Ready Estuaries (CRE) Programs:** Work with municipal and private water and wastewater utilities to promote use of the Climate Resilience, Evaluation and Awareness Tool (CREAT) to recognize and respond to climate change risks and promote with

National Estuary Program partners the use of the new Climate Ready Estuaries workbook to develop local climate vulnerability assessments and resilience plans.

8. **Develop Regional WaterSense Partners:** Work with States, Tribes, municipalities, non-profit organizations and businesses to promote the WaterSense Program in the region.

9. **Work with State Revolving Loan Fund Programs to recognize climate change impacts.**

IV. Office of Water Contribution to Meeting EPA Strategic Measures on Climate Change

The EPA *Strategic Plan* identifies several objectives for strengthening the Agency response to a changing climate by 2015. The Office of Water is committed to contributing to the Agency work to meet these objectives. The Agency objective and the Office of Water contribution to meeting the objective are identified below. Implementation plans from other offices within EPA address additional steps to be taken to meet these objectives.

- **Save energy and conserve resources:** The Office of Water will support this objective through measures to reduce energy use at wastewater treatment plants and through the WaterSense program.

- **Integrate climate change science into five major models and/or decision support tools:** The Office of Water supported this objective through publication of Version 2.0 of the Climate Resilience Evaluation and Awareness Tool (CREAT) for water utilities.

- **Integrate climate change science or trend information into five major rulemaking processes:** The Office of Water will support this objective through the development of a water program regulation prior to 2015. The specific regulation is not yet determined.

- **Integrate climate change considerations into five major grant, loan, or technical assistance programs:** The Office of Water is meeting this objective through integrating climate change in the National Estuary Program grant program.

The Office of Water will monitor progress in supporting these Agency objectives annually and will adjust programs and activities as needed to assure that the water program contributions to meeting the goals are achieved by the 2015 due date.

V. Legal and Enforcement Issues

The Office of Water works closely with the EPA Office of General Counsel and matters related to climate change and water resources and will continue this working relationship in the future. To date, water program actions to respond to a changing climate have not faced significant legal issues.

As noted in Section IX of this Plan, the Office of Water is interested in initiating a pilot project for collaboration with the EPA Office of Enforcement and Compliance Assurance addressing inclusion of climate change considerations in compliance and enforcement activities.

VI. Training and Outreach

The Office of Water will continue and expand current work to provide training to water program managers on climate change issues and to ensure strong communication and coordination among EPA water program offices, regional offices, other Federal agencies, and stakeholders. These training and outreach materials are addressed in Goal 17 of the *2012 Strategy*.

Training

The Office of Water will continue to work to provide training on climate change impacts on water resources, and especially on the impacts on clean water and drinking water programs. Some key actions the Office of Water will take include:

➢ Update the **Climate Change and Water training module** included as part of the EPA Watershed Academy online training program and promote this climate change and water training with EPA water program staff in national and regional offices;

➢ Continue the monthly **Climate Change and Water Seminar Series** which invites experts in climate change and water issues to speak to EPA Headquarters and Regional staff;

➢ Continue to support the **Climate Change Module** at the EPA Water Quality Standards Academy that presents an overview of climate change impacts on water resources and climate change vulnerability considerations for managers;

➢ Participate in the Agency workgroup tasked with **developing a general training tool for EPA staff addressing climate change adaptation challenges**;

➢ Work with Regions to develop a **model presentation** that EPA regional water programs can use to describe the climate change and water issues generally, with a focus on the implementation of the *National Water Program 2012 Strategy: Response to Climate Change;* and

➢ Sponsor an **all-hands meeting** of the Office of Water after the completion of the EPA *Climate Change Adaptation Implementation Plan* to generally familiarize all staff with the challenges that climate change poses and to describe the new Agency Implementation Plan with special emphasis on the *2012 Strategy* for the National Water Program and this Office of Water *Implementation Plan*.

Outreach

The Office of Water will support several activities to cooperate with other EPA offices, Federal agencies, and other organizations interested in addressing the impacts of a changing climate on water resources including:

➤ Continue to support the **National Water Program Climate Change Workgroup and provide oversight and management of National Water Program climate change actions,** including implementation of the *National Water Program 2012 Strategy: Response to Climate Change* (this Workgroup includes staff from national program offices, other EPA offices, EPA regional offices, and Great Waterbody offices);

➤ Continue to support the EPA National Water Program **State and Tribal Climate Change Council** that advises the National Water Program on a full range of climate change adaptation issues, including implementation of the *National Water Program 2012 Strategy: Response to Climate Change,* and engaging additional stakeholders using a range of forums and mechanisms.

➤ Continue to serve as **co-chair of the Interagency Water Resources Workgroup** that supports the Interagency Council on Climate Resilience and Preparedness and provide staff support to this Workgroup that oversees the implementation of the *National Action Plan: Priorities for Managing Freshwater Resources in a Changing Climate;*

➤ Serve as the **Federal Agency co-chair of the Climate Change Workgroup of the Advisory Committee on Water Information** (ACWI) in cooperation with the non-Federal co-chair from the Water Environment Federation;

➤ Continue to manage the EPA climate change and water **website** providing information and materials on a range of climate change and water topics;

➤ Continue to publish the EPA **Climate Change and Water News electronic newsletter** and consider options to expand the number of subscribers (currently approximately 2,700 email addresses) with a goal of doubling that number;

➤ Work with EPA Regions to support efforts to link and **coordinate Office of Water climate change website content** with climate change content provided on regional water program websites;

➤ Work with the U.S. Army Corps of Engineers in the effective operation and management of the **Water Resources Toolbox website** which provides a one stop website of information on a range of water resources management issues, including climate change;

➤ Work with EPA Region 10, the National Oceanic and Atmospheric Administration, the State of Washington, and other interested agencies and Tribes on issues related to assessing water quality criteria relevant to ocean and coastal acidification (e.g., aragonite saturation state (calcium carbonate availability for calcareous organisms' shell building); and

➤ Work with the U.S. Department of Energy to accelerate progress in understanding and developing innovative technologies and processes that lead to improved management of both water resources and energy production including topics such as:

  o Integrated water resource management;
  o Water and energy efficiency and conservation;
  o Water quality;
  o Use and reuse of wastewater for power generation;
  o Emergency response and recovery; and
  o Thermoelectric generation.

VII. Partnerships with Tribes

EPA values its unique government-to-government relationship with Indian Tribes in planning and decision making. Supporting the development of capacity to adapt to climate change among Tribes is a priority for the EPA. Tribes are particularly vulnerable to the impacts of climate change due to the integral nature of the environment within their traditional lifeways and culture. There is a strong need to develop adaptation strategies that promote sustainability and reduce the impact of climate change on Tribes.

EPA engaged Tribes through a formal consultation process in the development of the Agency's *Climate Change Adaptation Plan*. Tribes identified some of the most pressing issues as erosion, temperature change, drought, and various changes in access to and quality of water. Tribes recommended a number of tools and strategies to address these issues, including improving access to data and information; supporting baseline research to better track the effects of climate change; developing community-level education and awareness materials; and providing financial and technical support. These collaborative efforts will benefit from the expertise provide by our tribal partners and the Traditional Ecological Knowledge (TEK) they possess. TEK is a valuable body of knowledge in assessing the current and future impacts of climate change and has been used by Tribes for millennia as a valuable tool to adapt to changing surroundings.

The Office of Water has a strong partnership with Tribes and works closely with Tribes in the implementation of clean water and drinking water programs. This partnership extends to work on issues relating to climate change and water.

The *2012 Strategy* addresses cooperation with Tribes on climate change and water matters in Goals 15 and 16. Some key objectives of these goals include:

❖ Strategic Action 47: Through formal consultation and other mechanisms, incorporate climate change as a key consideration in the revised National Water Program Tribal Strategy and subsequent implementation of Clean Water Act (CWA), Safe Drinking Water Act (SDWA), and other core programs;

❖ Strategic Action 48: Incorporate adaptation into tribal funding mechanisms, and collaborate with other EPA and federal funding programs to support sustainability and adaptation in tribal communities;

❖ Strategic Action 49: Collaborate to explore and develop climate change science, information, and tools for Tribes, and incorporate local knowledge; and

❖ Strategic Action 50: Collaborate to develop communication materials relevant for tribal uses and tribal audiences.

Some key actions that the Office of Water will implement to advance the goal of supporting Tribes in responding to the water-related impacts of climate change include:

- ➢ Continue to support the EPA National Water Program **State and Tribal Climate Change Council** as a vehicle for sharing information and hearing the views of Tribes on climate change issues;

- ➢ Include a presentation addressing tribal climate change recommendations at the next Office of Water **tribal water quality conference**, planned for 2015;

- ➢ Continue to support sustainability and adaptation in tribal communities in coordination with the EPA-Tribal Science Council (TSC), as the TSC implements its tribal science priorities for climate change and Traditional Ecological Knowledge (TEK) (e.g. the Office of Water will develop and incorporate **water-resource specific information into materials, presentations, and training related to TEK);**

- ➢ Collaborate with the EPA American Indian Environmental Office (AIEO) to **incorporate climate change adaptation into the tribal General Assistance Program (GAP) grant guidance;**

- ➢ Work with Tribes to identify the most **pressing and significant impacts that a changing climate poses for tribal management of water resources** and support actions to respond to climate change related vulnerabilities; and

- ➢ Partner with tribal stakeholders to develop and pilot the **Tribal-Focused Environmental Risk Screening Tool (Tribal-FERST)**, a web-based geospatial and information access tool to support tribal environmental decision making that provides access to relevant science and information that can be used to help identify, prioritize, and manage environmental and public health issues.

VIII. Vulnerable Populations and Places

Certain parts of the population, such as children, the elderly, minorities, the poor, persons with underlying medical conditions and disabilities, those with limited access to information, and tribal and indigenous populations, can be especially vulnerable to the impacts of climate change. Also, certain geographic locations and communities are particularly vulnerable, such as those located in low-lying coastal areas. One of the principles guiding EPA's efforts to integrate climate adaptation into its programs, policies, and rules calls for its adaptation plans to prioritize helping people, places, and infrastructure that are most vulnerable to climate impacts and to be designed and implemented with meaningful involvement from all parts of society.

The Office of Water is giving special attention to populations and places that are most vulnerable to the water related impacts of a changing climate.   As noted above, the Office of Water is working closely with Tribes to respond to climate change impacts on water resources.

In the case of vulnerable places, the Office of Water is supporting national program initiatives that address places that are especially vulnerable to a changing climate. For example, the Climate Ready Estuaries program advances climate adaptation work in many of the 28 estuaries that participate in the National Estuary Program.

In addition, the Office of Water will work with EPA Regional offices to support climate change adaptation work by Great Waterbody offices (e.g., the Chesapeake Bay Program Office and the Great Lakes National Program Office) as well as other large ecosystem programs. Some examples of actions planned in this area are provided below.

   ➤ The Chesapeake Bay Program Office will develop a **research coordination and support program to address climate change issues in the Chesapeake Bay.**

   ➤ The **Great Lakes National Program Office will initiate the Climate Change Impacts Annex Subcommittee** to the newly formed Great Lakes Executive Committee under the Great Lakes Water Quality Agreement with Canada and develop and implement a bi-national workplan to undertake activities over the next three years to fulfill the commitments in the annex.

   ➤ EPA Region 10 will address **climate change in grants to support protection and restoration of Puget Sound** consistent with the Puget Sound Action Agenda including:

      o conduct an erosion survey to evaluate sea level rise threat in San Juan County;
      o map habitat and infrastructure vulnerability in Puget Sound and restoration potential for reducing vulnerability;
      o Tribes and counties will incorporate climate change in their plans and/or analyses.

Puget Sound Grant partners include: Puget Sound Partnership, Friends of the San Juans, The Nature Conservancy, Snohomish County, Washington Department of Ecology, Samish Indian Nation, Swinomish Tribe, Nooksack Tribe, Suquamish Tribe, and Port Gamble Indian Commission.

➢ EPA Regions 1 and 2 will revise the **Lake Champlain TMDL,** including an analysis of potential effects of climate change on phosphorous loads to the Lake.

➢ EPA Regions 2 and 3 will support the **Partnership for the Delaware Estuary** efforts toward climate change adaption planning by expanding upon the work of the climate change adaptation plan of 2010 including:

  ○ creating a **living shorelines process document** that combines their knowledge of the Delaware Estuary Living Shorelines Initiative planning, installation, and outreach processes and best practices; and

  ○ continue recruiting communities to the **Weathering Change** program in which agencies work with the community to help them understand the weather-related changes that are beginning to happen in their community.

➢ EPA Regions 1 and 2 are supporting the **Long Island Sound** program in implementing the **"Sentinels of Climate Change: Coastal Indicators of Wildlife and Ecosystem Change"** project in Long Island Sound. The project will address several of the key climate change sentinels identified by the Sentinel Monitoring program, including the responses of critical and sensitive habitats, such as salt marsh and tidal flats, and how changes in these ecosystems impact the population and behavior patterns of key bird species inhabiting them.

➢ EPA Regions 1, 2, 3, and 4 are working with other Federal agencies (e.g., the U.S. Department of the Interior, National Oceanic and Atmospheric Administration, and Army Corps of Engineers) and States **to manage development of off-shore renewable energy facilities, including identify areas best suited for wind energy production**. EPA will have significant National Environmental Policy Act (NEPA) responsibilities once projects are proposed and, to a lesser degree, Clean Water Act and Clean Air Act permitting responsibilities.

IX. Evaluation and Cross-Office Pilot Projects

Work is underway within the Office of Water to develop strong program evaluation practices for assessing progress in responding to climate change impacts and to develop effective collaborations with other EPA national program offices are described below.

Evaluation of Progress

The National Water Program initiated a new process in 2012 to track progress in implementing climate change response programs based on assessing the stage or phase of development of efforts to implement each of the 19 major Goals identified in the *2012 Strategy*. Progress toward each of the 19 Goals was assessed by program staff in the context of one of seven phases of development. The seven developmental phases are:

1. **Initiation**; conduct a screening assessment of potential implications of climate change to mission, programs, and operations;

2. **Assessment**; conduct a broader review to understand how climate change affects the resources in question;

3. **Response Development**; identify changes necessary to continue to reach program mission and goals and develop initial action plan;

4. **Initial Implementation**; initiate actions in selected priority programs or projects

5. **Robust Implementation**; programs are underway and lessons learned are being applied to additional programs and projects;

6. **Mainstreaming;** climate is an embedded, component of the program; and

7. **Monitor Outcomes and Adaptive Management;** continue to monitor and integrate performance, new information, and lessons learned into programs and plans.

In the 2012 *Highlights of Progress* (see http://www.epa.gov/water/climatechange), the Office of Water identified the status of work on each of the Goals in the *2012 Strategy* (see Appendix 1) as of December 2012.   This 2012 baseline assessment has a total numeric value of 43 out of a total possible score of 133 (i.e., 19 Goals times a score of 7 for each action = 133). This combined score indicates that many actions are in the early stages of implementation. Future annual progress reports will identify the cumulative progress toward full implementation of the *2012 Strategy* in both narrative and numeric terms.

In the future, the Office of Water will work to develop metrics that assess the readiness of clean water programs in the face of a changing climate and the contribution that water programs make toward reducing releases of greenhouse gases (e.g., reducing water use which reduces energy use, or generating energy from wastewater treatment to lower carbon footprints of these facilities).

Cross-Organization Projects

The EPA Office of Water is engaged in two major cross-organization projects related to climate change adaptation:

> **Collaboration on National Estuaries Program**: The Office of Water, in collaboration with the EPA Office of Air and Radiation (OAR) has funded 37 projects with 23 National Estuary Programs (NEPs) in six EPA Regions through the Climate Ready Estuaries Program. In 2012, the program completed the first Climate Ready Water Utilities pilot project, held a lessons learned workshop with NEPs in EPA Region 1, held a joint stakeholder meeting with the National Oceanic and Atmospheric Administration, and promoted Fall 2011 king tide sea level rise education campaigns with 10 NEPs. In addition, two NEPs collaborated with the EPA Office of Research and Development to pilot test an expert elicitation approach to address climate change vulnerability assessments. In 2013, the Office of Water will continue to work with OAR to help National Estuary Programs respond to a changing climate.

> **Collaboration on Evaluation:** The Office of Water has undertaken a measurement and evaluation project through the EPA Office of Policy's Evaluation Support Division to guide implementation of *National Water Program 2012 Strategy: Response to Climate Change* (*2012 Strategy*). The purpose of this project is to:

> - develop a robust performance measurement approach for the *2012 Strategy; and*
> - identify lessons learned from previous climate change planning efforts that can inform implementation of the new strategy.

> The Office of Water views measurement in general, and this project specifically, as critical for the long-term success of the *2012 Strategy.*

> **Collaboration on Climate and Water Research:** The Office of Water has a longstanding collaboration with the EPA Office of Research and Development (ORD) to ensure that climate change issues are addressed to the extent possible in water research supported by the Agency.   In 2013, the Office of Water and ORD will begin quarterly meetings to review progress and set directions for research related to climate change and water and will organize research projects from different parts of ORD according to the specific goals identified in the *2012 Strategy.*

Potential future collaborations with other EPA Offices include:

➢ **Collaboration on Enforcement Issues:** The Office of Water is interested in working with the EPA Office of Enforcement and Compliance Assurance to identify opportunities to recognize the impacts of a changing climate on water resources in the context of compliance and enforcement activities and actions through the use of green infrastructure, climate resilient approaches, and other measures.

➢ **Collaboration on Storm Surge Screening Criteria:** The Office of Water is working with Regions 1, 2, 3, 4, and 6 to begin development of initial screening criteria that could be used to identify water and wastewater facilities that may be at risk from inundation as a result of a storm surge event comparable to that generated by Hurricane Sandy.

Page intentionally left blank

APPENDICES

Page intentionally left blank

Appendix 1:
Visions, Goals and Strategic Actions of the
National Water Program 2012 Strategy: Response to Climate Change
Including 2012 Baseline Assessment Scores

Visions and Goals	Strategic Actions (SA)	2012 Development Phase / Score
Infrastructure: In the face of a changing climate, resilient and adaptable drinking water, wastewater and stormwater utilities (water sector) ensure clean and safe water to protect the nation's public health and environment by making smart investment decisions to improve the sustainability of their infrastructure and operations and the communities they serve, while reducing greenhouse gas emissions through greater energy efficiency.		
Goal 1: Build the body of information and tools needed to incorporate climate change into planning and decision making.	SA1: Improve access to vetted climate and hydrological science, modeling, and assessment tools through the Climate Ready Water Utilities Initiative. SA2: Assist wastewater and water utilities to reduce greenhouse gas emissions and increase long-term sustainability with a combination of energy efficiency, co-generation, and increased use of renewable energy resources. SA3: Work with the States and public water systems, particularly small water systems, to identify and plan for climate change challenges to drinking water safety and to assist in meeting health based drinking water standards. SA4: Promote sustainable design approaches to provide for the long-term sustainability of infrastructure and operations.	Phase: Response Development Score: 3
Goal 2: Support Integrated Water Resources Management (IWRM) to sustainably manage water resources.	SA5: Understand and promote through technical assistance the use of water supply management strategies. SA6: Evaluate and provide technical assistance on the use of water demand management strategies. SA7: Increase cross-sector knowledge of water supply climate challenges and develop watershed specific information to inform decision making.	Phase: Assessment Score: 2

Visions and Goals	Strategic Actions	2012 Development Phase / Score
Watersheds & Wetlands: Watersheds are protected, maintained and restored to ensure climate resilience and to preserve the social and economic benefits they provide; and the nation's wetlands are maintained and improved using integrated approaches that recognize their inherent value as well as their role in reducing the impacts of climate change.		
Goal 3: Identify, protect, and maintain a network of healthy watersheds and supportive habitat corridor networks.	SA8: Develop a national framework and support efforts to protect remaining healthy watersheds and aquatic ecosystems.	Phase: Response Development
	SA9: Collaborate with partners on terrestrial ecosystems and hydrology so that effects on water quality and aquatic ecosystems are considered.	
	SA10: Integrate protection of healthy watersheds throughout the National Water Program core programs.	Score: 3
	SA11: Increase public awareness of the role and importance of healthy watersheds in reducing the impacts of climate change.	
Goal 4: Incorporate climate resilience into watershed restoration and floodplain management.	SA12: Consider a means of accounting for climate change in EPA funded and other watershed restoration projects.	Phase: Response Development
	SA13: Work with federal, state, interstate, tribal, and local partners to protect and restore the natural resources and functions of riverine and coastal floodplains as a means of building resiliency and protecting water quality.	Score: 3
Goal 5: Watershed protection practices incorporate Source Water Protection to protect drinking water supplies.	SA14: Encourage States to update their source water delineations, assessments or protection plans to address anticipated climate change impacts.	Phase: Assessment
	SA15: Continue to support collaborative efforts to increase state and local awareness of source water protection needs and opportunities, and encourage inclusion of source water protection areas in local climate change adaptation initiatives.	Score: 2

Visions and Goals	Strategic Actions	2012 Baseline Assessment
Watersheds & Wetlands (continued)		
Goal 6: Incorporate climate change considerations into the Clean Water Act (CWA) 404 regulatory program as they relate to permit reviews and compensatory mitigation.	SA16: Consider the effects of climate change, as appropriate, when making significant degradation determinations in the CWA Section 404 wetlands permitting and enforcement program	Phase: Initiation Score: 1
	SA17: Evaluate, in conjunction with the U.S. Army Corps of Engineers, how wetland and stream compensation projects could be selected, designed, and sited to aid in reducing the effects of climate change.	
Goal 7: Improve baseline information on wetland extent, condition and performance to inform effective adaptation to climate change.	SA18: Expand wetland mapping by supporting wetland mapping coalitions and training on use of the new federal Wetland Mapping Standard.	Phase: Initiation Score: 1
	SA19: Produce a statistically valid, ecological condition assessment of the nation's wetlands.	
	SA20: Work with partners and stakeholders to develop information and tools to support long term planning and priority setting for wetland restoration projects.	

Visions and Goals	Strategic Actions	2012 Baseline Assessment
Coastal and Ocean Waters: Adverse effects of climate change and unintended adverse consequences of responses to climate change have been successfully prevented or reduced in the ocean and coastal environment. Federal, tribal, state, and local agencies, organizations, and institutions are working cooperatively; and information necessary to integrate climate change considerations into ocean and coastal management is produced, readily available, and used.		
Goal 8: Collaborate to ensure information and methodologies for ocean and coastal areas are collected, produced, analyzed, and easily available.	SA21: Collaborate to ensure that synergy occurs, lessons learned are transferred, federal efforts effectively help local communities, and efforts are not duplicative or at cross-purposes. SA22: Work within EPA and with the U.S. Global Change Research Program and other federal, tribal, and state agencies to collect, produce, analyze, and format knowledge and information needed to protect ocean and coastal areas and make it easily available.	Phase: Response Development Score: 3
Goal 9: EPA geographically targeted programs support and build networks of local, tribal, state, regional and federal collaborators to take effective adaptation measures for coastal and ocean environments.	SA23: Work with the National Water Program's larger geographic programs to incorporate climate change considerations, focusing on both the natural and built environments. SA24: Address climate change adaptation and build stakeholder capacity when implementing National Estuary Program Comprehensive Conservation and Management Plans and through the Climate Ready Estuaries Program. SA25: Conduct outreach and education, and provide technical assistance to state and local watershed organizations and communities to build adaptive capacity in coastal areas outside the National Estuary Program and Large Aquatic Ecosystem programs.	Phase: Assessment Score: 2

Visions and Goals	Strategic Actions	2012 Baseline Assessment
Coastal and Ocean Waters (continued)		
Goal 10: Address climate driven environmental changes in coastal areas and ensure that mitigation and adaptation are conducted in an environmentally responsible manner.	SA26: Support coastal wastewater, stormwater, and drinking water infrastructure owners and operators in reducing climate risks and encourage adaptation in coastal areas.	Phase: Assessment Score: 2
	SA27: Support climate readiness of coastal communities, including hazard mitigation, pre-disaster planning, preparedness, and recovery efforts.	
	SA28: Support preparation and response planning for diverse impacts to coastal aquatic environments.	
Goal 11: Ocean environments are protected by EPA programs that incorporate shifting environmental conditions, and other emerging threats.	SA29: Consider climate change impacts on marine water quality in National Water Program ocean management authorities, policies, and programs.	Phase: Assessment Score: 2
	SA30: Use available authorities and work with the regional ocean organizations and other federal and state agencies through regional ocean groups and other networks so that offshore renewable energy production does not adversely affect the marine environment.	
	SA31: Support the evaluation of sub-seabed sequestration of carbon dioxide (CO_2) and any proposals for ocean fertilization.	
	SA32: Participate in interagency development and implementation of federal strategies through the National Ocean Council and the National Ocean Council Strategic Action Plans.	

Visions and Goals	Strategic Actions	2012 Baseline Assessment
Water Quality: Our Nation's surface water, drinking water, and ground water quality are protected, and the risks of climate change to human health and the environment are diminished, through a variety of adaptation and mitigation strategies.		
Goal 12: Protect waters of the United States and promote management of sustainable surface water resources.	SA33: Encourage States and communities to incorporate climate change considerations into their water quality planning.	Phase: Assessment Score: 2
	SA34: Encourage green infrastructure and low-impact development to protect water quality and make watersheds more resilient.	
	SA35: Promote consideration of climate change impacts by National Pollutant Discharge Elimination System (NPDES) permitting authorities.	
	SA36: Encourage water quality authorities to consider climate change impacts when developing wasteload and load allocations in Total Maximum Daily Loads (TMDLs) where appropriate.	
	SA37: Identify and protect designated uses that are at risk from climate change impacts.	
	SA38: Clarify how to re-evaluate aquatic life water quality criteria on more regular intervals; and develop information to assist States and Tribes who are developing criteria that incorporate climate change considerations for hydrologic condition.	
Goal 13: As the Nation makes decisions to reduce its greenhouse gas emissions and develop alternative sources of energy and fuel, the National Water Program will work to protect water resources from unintended adverse consequences.	SA39: Continue to provide perspective on the water resource implications of new energy technologies.	Phase: Initiation Score: 1
	SA40: Provide assistance to states and permittees to assure that geologic sequestration of carbon dioxide is responsibly managed.	
	SA41: Continue to work with States to help them identify polluted waters, including those affected by biofuels production, and help them develop and implement Total Maximum Daily Loads (TMDLs) for those waters.	
	SA42: Provide informational materials for stakeholders to encourage the consideration of alternative sources of energy and fuels that are water efficient and maintain water quality.	
	SA43: As climate change affects the operation or placement of reservoirs, EPA will work with other Federal agencies and EPA programs to understand the combined effects of climate change and hydropower on flows, water temperature, and water quality.	

Visions and Goals	Strategic Actions	2012 Baseline Assessment
Water Quality (continued)		
Goal 14: Collaborate to make hydrological and climate data and projections available.	SA44: Monitor climate change impacts to surface waters and ground water.	Phase: Response Development
	SA45: Collaborate with other Federal agencies to develop new methods for use of updated precipitation, storm frequency, and observational streamflow data, as well as methods for evaluating projected changes in low flow conditions.	
	SA46: Enhance flow estimation using National Hydrography Dataset Plus (NHDPlus).	Score: 3
Working With Tribes: Tribes are able to preserve, adapt, and maintain the viability of their culture, traditions, natural resources, and economies in the face of a changing climate.		
Goal 15: Incorporate climate change considerations in the implementation of core programs, and collaborate with other EPA offices and Federal agencies to work with Tribes on climate change issues on a multi-media basis.	SA47: Through formal consultation and other mechanisms, incorporate climate change as a key consideration in the revised National Water Program Tribal Strategy and subsequent implementation of the Clean Water Act, Safe Drinking Water Act, and other core programs.	Phase: Assessment
	SA48: Incorporate adaptation into tribal funding mechanisms, and collaborate with other EPA and Federal funding programs to support sustainability and adaptation in tribal communities.	Score: 2
Goal 16: Tribes have access to information on climate change for decision making.	SA49: Collaborate to explore and develop climate change science, information, and tools for Tribes, and incorporate local knowledge.	Phase: Assessment
	SA50: Collaborate to develop communication materials relevant for tribal uses and tribal audiences.	Score: 2

Visions and Goals	Strategic Actions	2012 Baseline Assessment
Cross-Cutting Program Support		
Goal 17: Communication, Collaboration, and Training	SA51: Continue building the communication, collaboration, and training mechanisms needed to effectively increase adaptive capacity at the federal, tribal, state, and local levels.	Phase: Response Development Score: 3
Goal 18: Tracking Progress And Measuring Outcomes	SA52: Adopt a phased approach to track programmatic progress towards Strategic Actions; achieve commitments reflected in the Agency *Strategic Plan*; work with the National Water Program Climate Change Workgroup to develop outcome measures.	Phase: Response Development Score: 3
Goal 19: Climate Change and Water Research Needs	SA53: Work with the EPA Office of Research and Development, other water science agencies, and the water research community to further define needs and develop research opportunities to deliver the information needed to support implementation of this *2012 Strategy*, including providing the decision support tools needed by water resource managers.	Phase: Assessment Score: 2
		Total Score: **42 of a possible 133**

APPENDIX 2:
National Water Program Climate Change Organizational Framework

Assistant Administrator for Water

(Deputy Assistant Administrator chairs National Water Program Climate Change Workgroup)

EPA Planning and Budget

FY2011-2015 EPA Strategic Plan (Climate Change Adaptation Strategic Measures)

Council on Climate, Preparedness, and Reslience

Water Resources Workgroup EPA Office of Water, Co-chair

National Action Plan: Priorities for Managing Freshwater Resources in a Changing Climate

National Water Program Climate Change Workgroup

(EPA Headquarters Water Program Offices and Regions)

2012 National Water Program Climate Change Strategy

National Water Program Climate Change Adaptation Workplan (Internal)

National Water Program Climate Change Adaptation Implementation Plan (part of Agency Climate Plan)

National Water Program Planning and Budget

EPA National Water Program Guidance (Climate Section)

Interagency Advisory Committee on Water Information (ACWI)

Water Resources and Climate Change Workgroup (WRCCWG)

EPA Office of Water, Federal Co-chair

EPA Cross-Agency Adaptation Workgroup

EPA Climate Change Adaptation Plan (June 2012)

EPA Program Office/Region-Specific Climate Adaptation Implementation Plans

EPA 10 Regional Offices

Climate Change Adaptation Implementation Plans (part of Agency Climate Change Implementation Plans)

EPA National Water Program

State and Tribal Climate Change Council

Appendix 3:
National Water Program Climate Change Workgroup
Principal Members

Office of Ground Water and Drinking Water
Mike Muse
Curt Baranowski
Office of Science and Technology
Rachael Novak
Office of Wastewater Management
Veronica Blette
Karen Metchis
Lynn Stabenfeld
Office of Wetlands, Oceans and Watersheds
Michael Craghan
Kathleen Kutschenreuter
Julie Reichert
Office of the Assistant Administrator for Water
Mike Shapiro
Jeff Peterson
Elana Goldstein
David Bylsma
Regions
Region 1 - Mel Cote
Region 2 - Alexandre Remnek
Region 3 - Joe Piotrowski
Region 4 - Bob Howard
Region 5 - Kate Balasa
Region 6 - Jim Brown
Region 7 - Mary Mindrup
Region 8 - Mitra Jha
Region 9 - Suzanne Marr
Region 10 - Paula VanHaagen

Appendix 4: Table Illustrating Climate Change Impacts on Clean Water and Drinking Water Program Components

Clean Water/ Drinking Water Program Component	Climate Change Impacts					
	Increased Water Pollution	More Extreme Weather Events	Changes in Water Availability	Sea Level Rise	Coastal Area Impacts	Water Impacts of Energy Production
Drinking Water Standards	■					
Drinking Water Planning	■	■	■	■		■
Underground Injection Control Permits			■			■
Source Water Protection	■	■	■			■
Drinking Water SRF (State Revolving Funds)	■		■			
Surface Water Standards	■		■			■
Clean Water Planning	■	■	■	■		■
Discharge Permits	■	■	■			■
Nonpoint Pollution Control	■	■	■			■
Clean Water SRF (State Revolving Funds)	■	■	■			■
Technology Based Standards						■
Water Monitoring		■			■	
Storm Water Permits		■	■			
Coastal Zone	■	■		■	■	
Ocean Protection	■	■		■	■	■
Emergency Planning	■	■		■	■	■
Water Restoration/ Total Maximum Daily Loads (TMDLs)	■	■	■		■	■
Wetlands Permits	■	■		■	■	
National Estuaries Program	■	■		■	■	
Combined Sewer Overflow Plans	■	■				

www.ingramcontent.com/pod-product-compliance
Lightning Source LLC
Chambersburg PA
CBHW080651180526
45168CB00008B/3376